Editor
Lorin E. Klistoff, M.A.

Managing Editor
Karen Goldfluss, M.S. Ed.

Editor-in-Chief
Sharon Coan, M.S. Ed.

Cover Artist
Barb Lorseyedi

Art Coordinator
Kevin Barnes

Art Director
CJae Froshay

Imaging
Rosa C. See

Product Manager
Phil Garcia

Publishers
Rachelle Cracchiolo, M.S. Ed.
Mary Dupuy Smith, M.S. Ed.

Practice Makes Perfect

Math Review

GRADE 2

Author

Mary Rosenberg

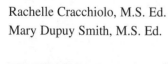

Teacher Created Materials, Inc.
6421 Industry Way
Westminster, CA 92683
www.teachercreated.com
ISBN-0-7439-3742-2
©2003 Teacher Created Materials, Inc.
Made in U.S.A.

Table of Contents

Introduction

The old adage "practice makes perfect" can really hold true for your child and his or her education. The more practice and exposure your child has with concepts being taught in school, the more success he or she is likely to find. For many parents, knowing how to help your children can be frustrating because the resources may not be readily available. As a parent it is also difficult to know where to focus your efforts so that the extra practice your child receives at home supports what he or she is learning in school.

This book has been designed to help parents and teachers reinforce basic skills with children. *Practice Makes Perfect* reviews basic math skills for children in grade 2. The focus is a review of math skills. While it would be impossible to include all concepts taught in grade 2 in this book, the following basic objectives are reinforced through practice exercises. These objectives support math standards established on a district, state, or national level. (Refer to the Table of Contents for specific objectives of each practice page.)

- adding and subtracting
- multiplying and dividing
- adding and subtracting money
- using a number line
- using a calendar

- using fractions
- using standard form
- finding the perimeter
- finding important information
- telling time

- measuring in centimeters
- counting by 3s
- beginning division
- identifying shapes

There are 36 practice pages. (*Note*: Have children show all work where computation is necessary to solve a problem. For multiple choice responses on practice pages, children can fill in the letter choice or circle the answer.) Following the practice pages are six test practices. These provide children with multiple-choice test items to help prepare them for standardized tests administered in schools. As your child completes each test, he or she can fill in the correct bubbles on the optional answer sheet provided on page 46. To correct the test pages and the practice pages in this book, use the answer key provided on pages 47 and 48.

How to Make the Most of This Book

Here are some useful ideas for optimizing the practice pages in this book:

- Set aside a specific place in your home to work on the practice pages. Keep it neat and tidy with materials on hand.

- Set up a certain time of day to work on the practice pages. This will establish consistency. Look for times in your day or week that are less hectic and more conducive to practicing skills.

- Keep all practice sessions with your child positive and constructive. If the mood becomes tense or you and your child are frustrated, set the book aside and look for another time to practice with your child.

- Help with instructions if necessary. If your child is having difficulty understanding what to do or how to get started, work through the first problem with him or her.

- Review the work your child has done. This serves as reinforcement and provides further practice.

- Allow your child to use whatever writing instruments he or she prefers. For example, colored pencils can add variety and pleasure to drill work.

- Pay attention to the areas in which your child has the most difficulty. Provide extra guidance and exercises in those areas. Allowing children to use drawings and manipulatives, such as coins, tiles, game markers, or flash cards, can help them grasp difficult concepts more easily.

- Look for ways to make real-life applications to the skills being reinforced.

Practice 1 ᗒ ᗰ ᗒ ᗰ ᗒ ᗰ ᗒ ᗰ ᗒ ᗰ ᗒ ᗰ ᗒ ᗰ ᗒ ᗰ

Use the number line to help solve the word problems.

---

1. Selina counted 5 cars. Dana counted 4 more than Selina. How many cars did Dana count in all?

5   6   7   8   9   10   11

Dana counted _____ cars in all.

2. Billy counted 8 fish in the koi pond. Jimmy counted 6 more fish than Billy. How many fish did Jimmy count in all?

8   9   10   11   12   13   14

Jimmy counted _____ fish in all.

---

3. Joey saw 14 tables in the picnic areas. Five of the tables were already being used. How many tables are left?

8   9   10   11   12   13   14

There are _____ tables left.

4. Zoe runs 15 miles over the weekend. On Saturday she ran 5 miles. How many miles does she still need to run?

9   10   11   12   13   14   15

Zoe needs to run _____ miles.

---

5. Danielle picked 12 carrots from her garden. Nate picked 6 more carrots than Danielle. How many carrots did Nate pick in all?

12   13   14   15   16   17   18

Nate picked _____ carrots in all.

6. Celeste bought 19 seed packets. She used 3 packets when planting the first row. How many packets does she have left?

13   14   15   16   17   18   19

Celeste has _____ packets left.

Write the sign (+ or −) to make each math sentence true.

7. 3 ____ 5 = 8   **8.** 5 ____ 3 = 8   **9.** 8 ____ 3 = 5

10. 8 ____ 5 = 3   **11.** 6 ____ 6 = 12   **12.** 12 ____ 6 = 6

Practice 2

Solve each problem.

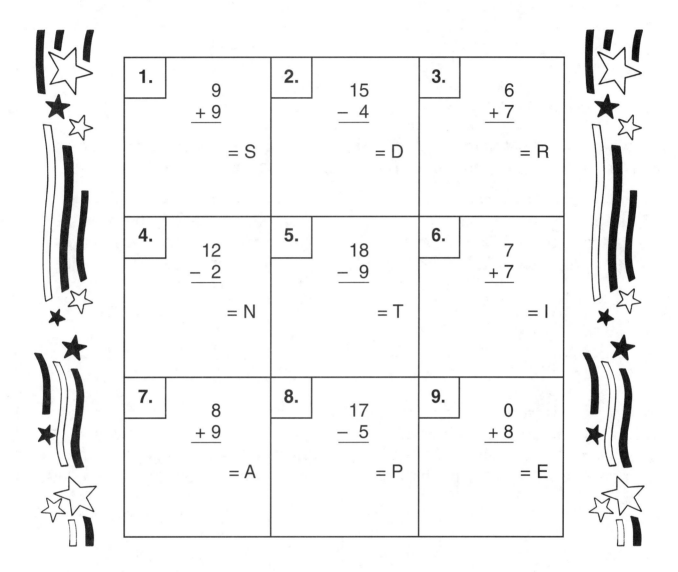

1. $\begin{array}{r} 9 \\ +\ 9 \\ \hline \end{array}$ = S	**2.** $\begin{array}{r} 15 \\ -\ 4 \\ \hline \end{array}$ = D	**3.** $\begin{array}{r} 6 \\ +\ 7 \\ \hline \end{array}$ = R
4. $\begin{array}{r} 12 \\ -\ 2 \\ \hline \end{array}$ = N	**5.** $\begin{array}{r} 18 \\ -\ 9 \\ \hline \end{array}$ = T	**6.** $\begin{array}{r} 7 \\ +\ 7 \\ \hline \end{array}$ = I
7. $\begin{array}{r} 8 \\ +\ 9 \\ \hline \end{array}$ = A	**8.** $\begin{array}{r} 17 \\ -\ 5 \\ \hline \end{array}$ = P	**9.** $\begin{array}{r} 0 \\ +\ 8 \\ \hline \end{array}$ = E

Write the letter on the line that matches each answer above.

___	___	___	___	___		___	___	___
18	9	17	13	18		17	10	11

___	___	___	___	___	___	___
18	9	13	14	12	8	18

Practice 3   ৹ ☙ ৹ ☙ ৹ ☙ ৹ ☙ ৹ ☙ ৹ ☙ ৹ ৹ ☙

The words below are often used in word problems and tell us what to do. On the line, write + if you should add or − if you should subtract.

1. in all _____   **4.** difference _____

2. fewer _____   **5.** were left _____

3. all together _____   **6.** sum _____

Read each word problem. Underline the phrase that tells whether it is an addition or subtraction problem. Then circle the correct math sentence.

Example: Sam made 3 origami birds and 4 origami fish. How many origami animals did Sam make <u>in all</u>? (3 + 4 = 7) 7 − 4 = 3	**7.** Meredith made 5 paper walkie-talkies and 5 paper hats. How many paper items did Meredith make in all? 5 + 5 = 10 10 − 5 = 5
8. Eric collected 9 shiny pebbles. He gave his little brother 6 of the pebbles. How many pebbles does Eric have left? 3 + 6 = 9 9 − 6 = 3	**9.** Tanya saw 2 lion kites and 7 dragon kites. Find the difference between the lion and dragon kites. 7 + 2 = 9 7 − 2 = 5
10. Bert planted 1 oak tree. Oscar planted 8 oak trees. How many trees did they plant all together? 1 + 8 = 9 9 − 1 = 8	**11.** Judy counted 7 birds. If 5 of them flew away, how many fewer birds were there? 5 + 7 = 12 7 − 5 = 2

Practice 4 ꩜ ꩜ ꩜ ꩜ ꩜ ꩜ ꩜ ꩜ ꩜

Add.

1. 87 + 9	2. 33 + 17	3. 64 + 26	4. 13 + 49	5. 25 + 25	6. 26 + 68

Use the chart to answer the questions. Fill in the correct circle.

Player	Friday	Saturday
Benita	54	26
Ken	68	14
Forest	36	48

7. Who scored the most points at Friday's game?

Benita   Forest   Ken

○   ○   ○

8. Who scored the most points at Saturday's game?

Benita   Forest   Ken

○   ○   ○

9. What was the total number of points scored by Benita?

70   80   90

○   ○   ○

10. What was the total number of points scored by Forest?

78   76   84

○   ○   ○

Practice 5

Each section of the bar equals one centimeter (cm). Measure each item to the nearest centimeter.

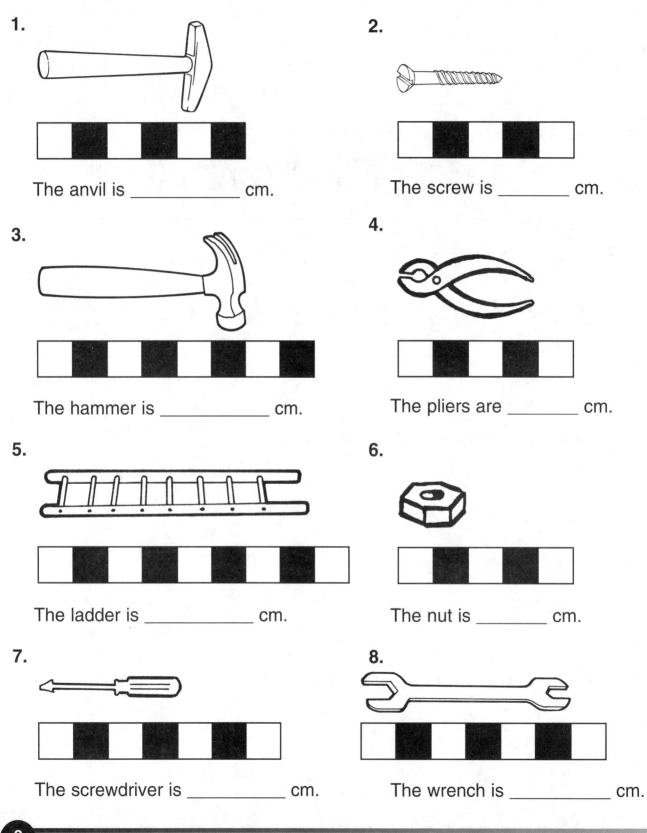

1.

The anvil is _____ cm.

2.

The screw is _____ cm.

3.

The hammer is _____ cm.

4.

The pliers are _____ cm.

5.

The ladder is _____ cm.

6.

The nut is _____ cm.

7.

The screwdriver is _____ cm.

8.

The wrench is _____ cm.

#3742 Practice Makes Perfect: Math Review

Practice 6

Count each shape's sides and corners.

1. Parallelogram	2. Octagon	3. Pentagon	4. Triangle
Sides:_____	Sides:_____	Sides:_____	Sides:_____
Corners:_____	Corners:_____	Corners:_____	Corners:_____

Read the clues to find each shape. Fill in the circle under the correct answer.

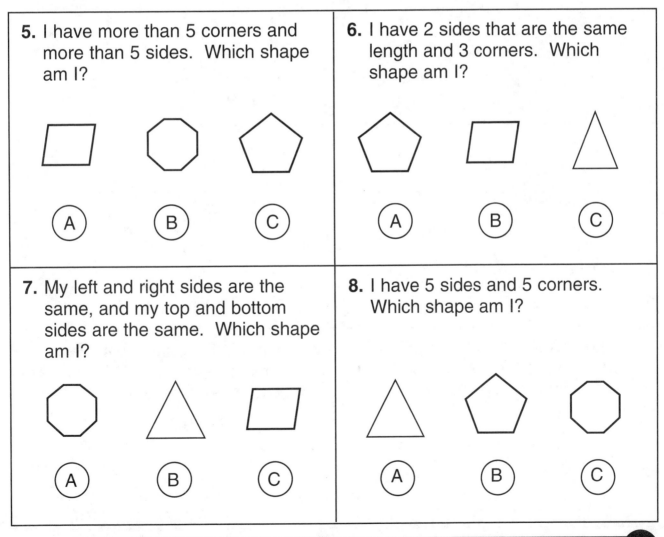

5. I have more than 5 corners and more than 5 sides. Which shape am I?

(A)   (B)   (C)

6. I have 2 sides that are the same length and 3 corners. Which shape am I?

(A)   (B)   (C)

7. My left and right sides are the same, and my top and bottom sides are the same. Which shape am I?

(A)   (B)   (C)

8. I have 5 sides and 5 corners. Which shape am I?

(A)   (B)   (C)

Practice 7

Find the perimeter of each shape.

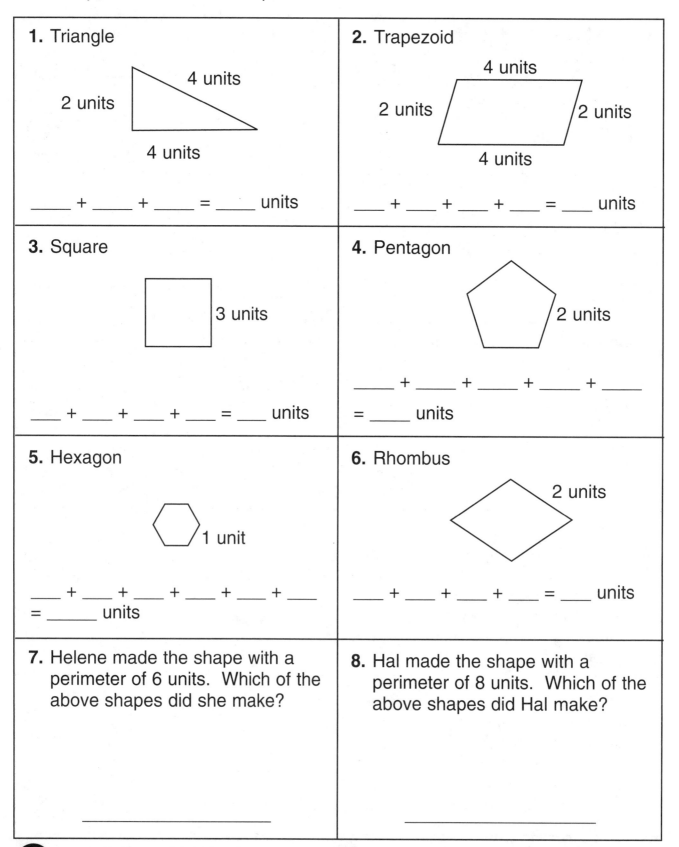

1. Triangle

4 units

2 units

4 units

____ + ____ + ____ = ____ units

2. Trapezoid

4 units

2 units   2 units

4 units

___ + ___ + ___ + ___ = ___ units

3. Square

3 units

___ + ___ + ___ + ___ = ___ units

4. Pentagon

2 units

____ + ____ + ____ + ____ + ____

= ____ units

5. Hexagon

1 unit

___ + ___ + ___ + ___ + ___ + ___
= _____ units

6. Rhombus

2 units

___ + ___ + ___ + ___ = ___ units

7. Helene made the shape with a perimeter of 6 units. Which of the above shapes did she make?

_____

8. Hal made the shape with a perimeter of 8 units. Which of the above shapes did Hal make?

_____

Practice 8

Color the coins that make the correct amount.

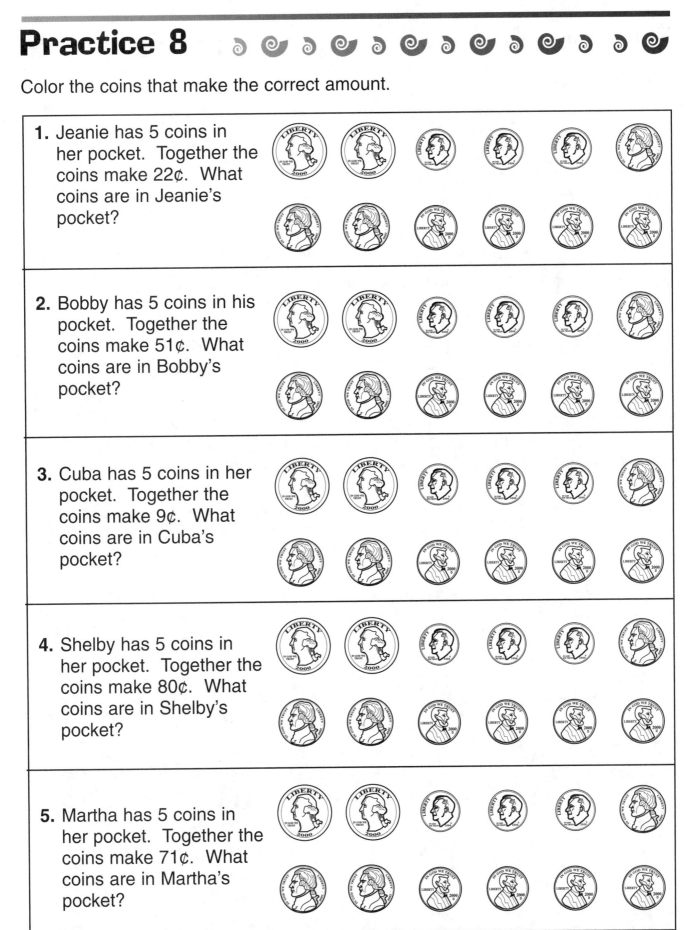

1. Jeanie has 5 coins in her pocket. Together the coins make 22¢. What coins are in Jeanie's pocket?

2. Bobby has 5 coins in his pocket. Together the coins make 51¢. What coins are in Bobby's pocket?

3. Cuba has 5 coins in her pocket. Together the coins make 9¢. What coins are in Cuba's pocket?

4. Shelby has 5 coins in her pocket. Together the coins make 80¢. What coins are in Shelby's pocket?

5. Martha has 5 coins in her pocket. Together the coins make 71¢. What coins are in Martha's pocket?

Practice 9

Finish the pattern.

1. 25, 50, _____ , 100, _____ , _____ , _____ , 200, _____

Fill in the circle under the correct amount.

2. Tyra has 5 quarters. How much money does she have?	**3.** Larry has 3 quarters. How much money does he have?

2.
$0.25 $1.05 $1.25
○   ○   ○

3.
$0.75 $1.75 $3.70
○   ○   ○

4. Cara has 1 quarter. How much money does she have?

$0.01 $0.05 $0.25
○   ○   ○

5. Ed has 6 quarters. How much money does he have?

$0.30 $1.50 $6.00
○   ○   ○

6. Tammy has 4 quarters. How much money does she have?

$0.01 $0.10 $1.00
○   ○   ○

7. Max has 7 quarters. How much money does he have?

$1.25 $1.75 $2.00
○   ○   ○

Practice 10 ❧ ❧ ❧ ❧ ❧ ❧ ❧ ❧ ❧ ❧ ❧ ❧ ❧ ❧

Count the money and write how much each item costs on the line.

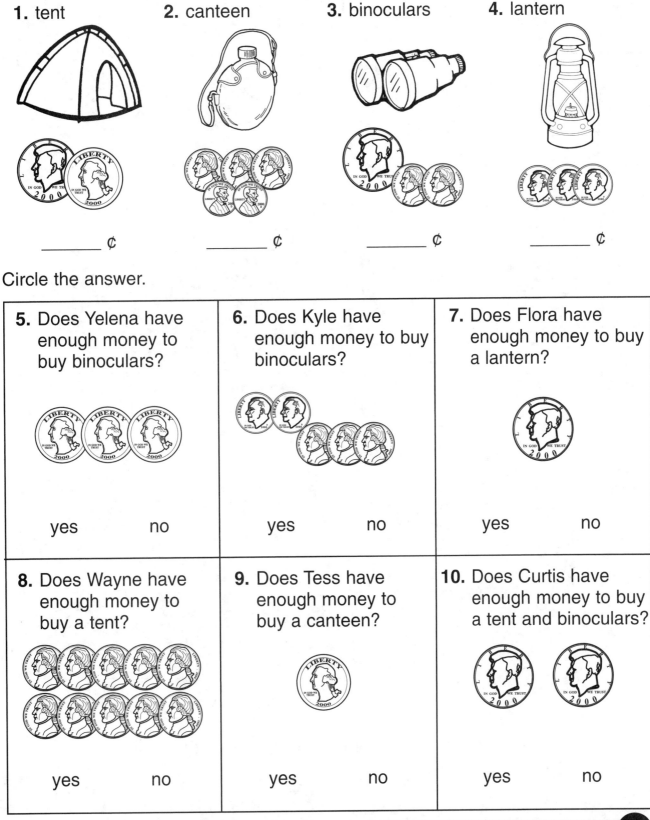

1. tent

2. canteen

3. binoculars

4. lantern

_____ ¢

_____ ¢

_____ ¢

_____ ¢

Circle the answer.

5. Does Yelena have enough money to buy binoculars? yes no	**6.** Does Kyle have enough money to buy binoculars? yes no	**7.** Does Flora have enough money to buy a lantern? yes no
8. Does Wayne have enough money to buy a tent? yes no	**9.** Does Tess have enough money to buy a canteen? yes no	**10.** Does Curtis have enough money to buy a tent and binoculars? yes no

    #3742 Practice Makes Perfect: Math Review

Practice 11

Count the money the students found in their pockets.

1.

$ _____

2.

$ _____

3.

$ _____

4.

$ _____

5.

$ _____

6.

$ _____

7. Write the amounts of money in order from smallest to largest.

_____, _____, _____, _____, _____, _____

Practice 12 ✺ ✺ ✺ ✺ ✺ ✺ ✺ ✺ ✺ ✺ ✺

Write the subtraction problem. Add to check the answer.

$42.16	$51.37	$11.06	$10.48	$30.66	$24.79

Example: Rosa has $52.20. She buys an area rug. How much money does she have left?

Rosa has $10.04 left.

$52.20
− $42.16
————
$10.04

$10.04
+ $42.16
————
$52.20

---

1. Theo has $79.50. He buys a new couch. How much money does Theo have left?

Theo has _____ left.

$__ __ . __ __
− $__ __ . __ __
——————
$__ __ . __ __

$__ __ . __ __
+ $__ __ . __ __
——————
$__ __ . __ __

---

2. Queena has $13.10. She buys a chair. How much money does Queena have left?

Queena has _____ left.

$__ __ . __ __
− $__ __ . __ __
——————
$__ __ . __ __

$__ __ . __ __
+ $__ __ . __ __
——————
$__ __ . __ __

---

3. Thomas has $11.75. He buys a new lamp. How much money does Thomas have left?

Thomas has _____ left.

$__ __ . __ __
− $__ __ . __ __
——————
$__ __ . __ __

$__ __ . __ __
+ $__ __ . __ __
——————
$__ __ . __ __

Practice 13 ⟋ ◉ ✧ ◉ ✧ ◉ ✧ ◉ ✧ ◉ ✧ ◉ ✧ ◉ ⟋ ⟋ ◉

Write the subtraction problem. Add to check the answer.

$42.16	$51.37	$11.06	$10.48	$30.66	$24.79

1. Maurine has $38.95. She buys a new TV set. How much money does Maurine have left?

$\$\_\_ \_\_ . \_\_ \_\_$

$- \$\_\_ \_\_ . \_\_ \_\_$

$\$\_\_ \_\_ . \_\_ \_\_$

$\$\_\_ \_\_ . \_\_ \_\_$

$+ \$\_\_ \_\_ . \_\_ \_\_$

$\$\_\_ \_\_ . \_\_ \_\_$

Maurine has _____ left.

2. Monty has $86.93. He buys a new table. How much money does Monty have left?

$\$\_\_ \_\_ . \_\_ \_\_$

$- \$\_\_ \_\_ . \_\_ \_\_$

$\$\_\_ \_\_ . \_\_ \_\_$

$\$\_\_ \_\_ . \_\_ \_\_$

$+ \$\_\_ \_\_ . \_\_ \_\_$

$\$\_\_ \_\_ . \_\_ \_\_$

Monty has _____ left.

3. Kiva has $83.76 to spend. Make a list of what she could buy and how much money she will have left.

Kiva can buy _____ , _____ , _____ .

Kiva will have _____ left.

Use the > (greater than) or < (less than) symbols to compare the price of the items.

4. ◯   **5.** ◯   **6.** ◯

Practice 14 ◖ ◖ ◖ ◖ ◖ ◖ ◖ ◖ ◖ ◖ ◖ ◖

Complete the calendar.

December

Sunday	Monday	Tuesday	Wednesday	Thursday	Friday	Saturday
						1
2					7	
				13		
		18				
			26			
	31					

Answer the questions.

1. What is the name of the month? _____

2. How many days are in this month? _____

Write the day of the week for the following dates:

3. 24th _____   4. 1st _____

5. 19th _____   6. 25th _____

7. 3rd _____   8. 16th _____

Practice 15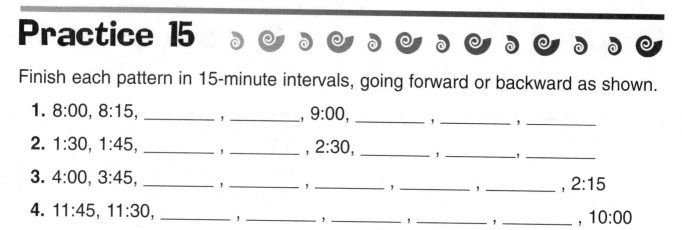

Finish each pattern in 15-minute intervals, going forward or backward as shown.

1. 8:00, 8:15, _____ , _____ , 9:00, _____ , _____ , _____

2. 1:30, 1:45, _____ , _____ , 2:30, _____ , _____ , _____

3. 4:00, 3:45, _____ , _____ , _____ , _____ , _____ , 2:15

4. 11:45, 11:30, _____ , _____ , _____ , _____ , _____ , 10:00

Draw the hands on the clock to show **15 minutes later** than the time given. Write the time on the lines below the clock.

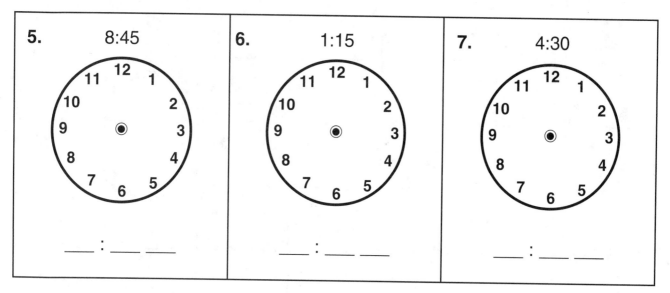

Draw the hands on the clock to show **15 minutes earlier** than the time given. Write the time on the lines below the clock.

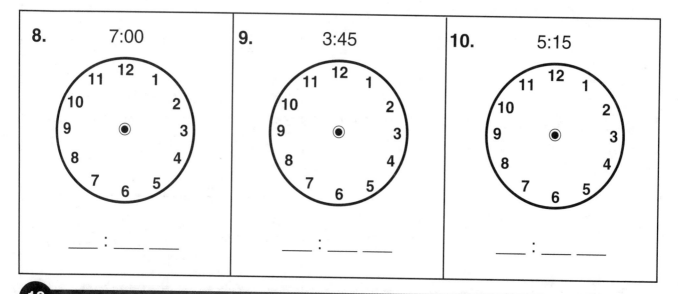

Practice 16 ୬ ୭ ୬ ୭ ୬ ୬ ୬ ୭ ୬ ୭ ୬ ୭ ୬ ୬ ୬ ୭

Solve each word problem.

---

1. The train was supposed to be here at 10:00. It arrived 15 minutes late. What time did it reach the station?

half past 10   a quarter past 10

2. The store usually opens at 9:00. Today though, it opened half an hour early. What time did the store open?

half past 8   a quarter past 8

---

3. Snack time is usually at 3:30. Today snack time was 15 minutes late. What time did the kids get to eat their snacks?

half past 4   a quarter till 4

4. Greg goes to bed at 7:00. Tonight Greg gets to stay up an extra half hour. What is Greg's bedtime for tonight?

half past 7   a quarter past 7

---

5. The movie was over at 8:15. There was a short 15 minutes cartoon after the movie. What time did the cartoon end?

half past 8   a quarter til 8

6. Max walks every day for half an hour. He left his house at 5:15. What time did he return back home?

half past 5   a quarter til 6

---

7. Frank put a cake in the oven at 6:15. It took half an hour to bake. What time did Frank take the cake out of the oven?

half past 7   a quarter til 7

8. It takes Molly half an hour to set up the volleyball net. If the game is to start at 2:00, what time should Molly start putting up the net?

half past 1   a quarter past 1

Practice 17 ⟨⟩ ⟨⟩ ⟨⟩ ⟨⟩ ⟨⟩ ⟨⟩ ⟨⟩ ⟨⟩ ⟨⟩

Match each picture to its estimate.

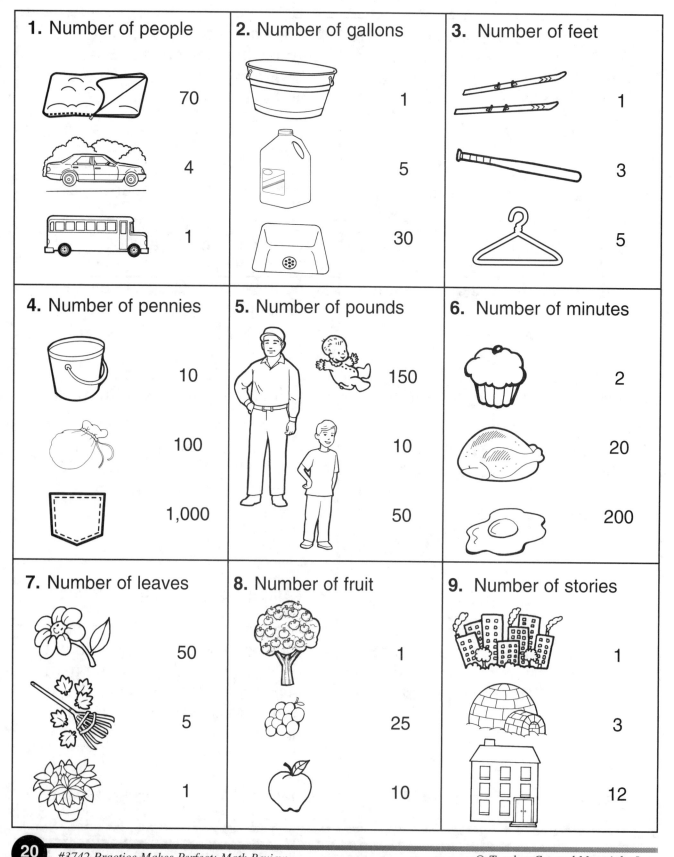

1. Number of people
70
4
1

2. Number of gallons
1
5
30

3. Number of feet
1
3
5

4. Number of pennies
10
100
1,000

5. Number of pounds
150
10
50

6. Number of minutes
2
20
200

7. Number of leaves
50
5
1

8. Number of fruit
1
25
10

9. Number of stories
1
3
12

Practice 18 ♪ ✲ ♪ ✲ ♪ ✲ ♪ ✲ ♪ ✲ ♪ ✲ ♪ ✲ ♪ ✲ ♪ ✲

Read and solve each word problem.

1. Reba counted 25 walnuts and 40 pecans. How many nuts did Reba count in all? ＿＿ ＿＿ +＿＿ ＿＿ ＿＿＿＿ ＿＿ ＿＿ Reba counted ＿＿＿ nuts in all.	**2.** Jordan counted 50 pistachios and 31 almonds. How many nuts did Jordan count in all? ＿＿ ＿＿ +＿＿ ＿＿ ＿＿＿＿ ＿＿ ＿＿ Jordan counted ＿＿＿ nuts in all.
3. Cecil counted 86 walnuts. He put 33 walnuts in a bag. How many walnuts are left? ＿＿ ＿＿ −＿＿ ＿＿ ＿＿＿＿ ＿＿ ＿＿ There are ＿＿＿ walnuts left.	**4.** Patsy counted 60 peanuts and 14 chestnuts. How many nuts did Patsy count in all? ＿＿ ＿＿ +＿＿ ＿＿ ＿＿＿＿ ＿＿ ＿＿ Patsy counted ＿＿＿ nuts in all.
5. Tyra counted 95 peanuts. She put 51 of them in a bag. How many peanuts are left? ＿＿ ＿＿ −＿＿ ＿＿ ＿＿＿＿ ＿＿ ＿＿ There are ＿＿＿ peanuts left.	**6.** Angel counted 73 pistachios. He put 22 of them in a bag. How many pistachios are left? ＿＿ ＿＿ −＿＿ ＿＿ ＿＿＿＿ ＿＿ ＿＿ There are ＿＿＿ pistachios left.

Practice 19

Rewrite each address in standard form.

Example 400 + 10 + 6 Apple Road 416 Apple Road _____	**1.** 300 + 30 + 4 Orange Street _____
2. 700 + 80 + 8 Pecan Avenue _____	**3.** 400 + 40 + 3 Walnut Lane _____
4. 900 + 10 + 1 Mandarin Lane _____	**5.** 600 + 10 + 9 Lemon Street _____
6. 600 + 80 + 2 Coconut Boulevard _____	**7.** Write your address in standard form. _____

Practice 20

Read the clues. Cross out the information that is not needed to solve the problem. Write each pet's name on the line.

Ralph	Dudley	Buster	Oscar
489	810	345	163

Lisa's Pet

1. Lisa's dog has a tag with 2 even numbers.

2. One even number is twice the other even number.

3. Lisa collects stamps in her free time.

Which dog is Lisa's?

_____

Pet

1 _____ dress like a cowboy.

_____ has a tag with two odd _____.

_____ hen the numbers in the tag are added together, they equal 12.

Which dog is Corey's?

_____

Ted's Pet

1. Ted's dog has a tag with an even number in it.

2. Ted likes to play the piano.

3. The tag has a 1 in it.

4. When the numbers are added together, they equal 9.

Which dog is Ted's?

_____

Anna's Pet

1. Anna's dog has a tag with two odd numbers.

2. One odd number is half that of one of the other numbers.

3. When Anna grows up, she would like to be a surfer.

Which dog is Anna's?

_____

Practice 21 ꙮ ꙮ ꙮ ꙮ ꙮ ꙮ ꙮ ꙮ ꙮ ꙮ ꙮ ꙮ

Solve each riddle. Fill in the correct circle.

1. Bay is thinking of a number. The number has a 1 in the hundreds place and a 2 in the ones place. What is Bay's number?

210   102   201
○   ○   ○

2. Mark is thinking of a number. The number has a 9 in the ones place and a 9 in the tens place. What is Mark's number?

999   957   369
○   ○   ○

3. Gage is thinking of a number. The number has a 7 in the hundreds place and a 5 in the tens place. What is Gage's number?

735   537   753
○   ○   ○

4. Brooke is thinking of a number. The number has a 7 in the tens place and a 4 in the ones place. What is Brooke's number?

476   674   764
○   ○   ○

5. Kate is thinking of a number. The number has a 3 in the hundreds place and a 5 in the ones place. What is Kate's number?

356   365   536
○   ○   ○

6. Kurt is thinking of a number. The number has a 4 in the hundreds place and an 8 in the tens place. What is Kurt's number?

284   482   842
○   ○   ○

Practice 22 🐚 🐚 🐚 🐚 🐚 🐚 🐚 🐚 🐚 🐚 🐚 🐚 🐚 🐚 🐚

Use the > (greater than) or < (less than) symbol. Then fill in the correct name on the line.

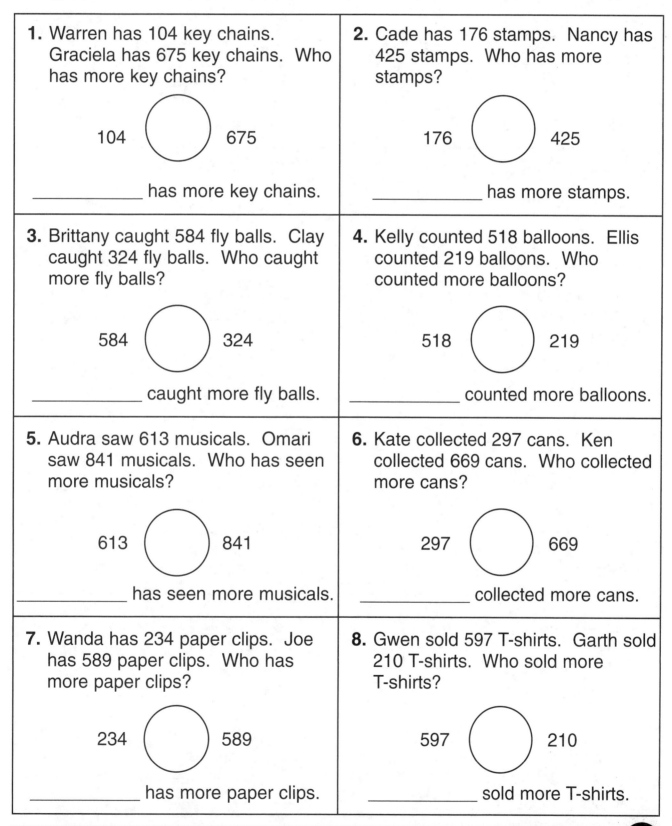

1. Warren has 104 key chains. Graciela has 675 key chains. Who has more key chains?

104 ◯ 675

_____ has more key chains.

2. Cade has 176 stamps. Nancy has 425 stamps. Who has more stamps?

176 ◯ 425

_____ has more stamps.

3. Brittany caught 584 fly balls. Clay caught 324 fly balls. Who caught more fly balls?

584 ◯ 324

_____ caught more fly balls.

4. Kelly counted 518 balloons. Ellis counted 219 balloons. Who counted more balloons?

518 ◯ 219

_____ counted more balloons.

5. Audra saw 613 musicals. Omari saw 841 musicals. Who has seen more musicals?

613 ◯ 841

_____ has seen more musicals.

6. Kate collected 297 cans. Ken collected 669 cans. Who collected more cans?

297 ◯ 669

_____ collected more cans.

7. Wanda has 234 paper clips. Joe has 589 paper clips. Who has more paper clips?

234 ◯ 589

_____ has more paper clips.

8. Gwen sold 597 T-shirts. Garth sold 210 T-shirts. Who sold more T-shirts?

597 ◯ 210

_____ sold more T-shirts.

Practice 23 🌀 🐚 🌀 🐚 🌀 🐚 🌀 🐚 🌀 🐚 🌀 🐚 🌀 🐚 🌀

Write and solve each math problem.

mittens	**skis**	**hats**	**jackets**	**boots**	**pants**
403	230	512	154	125	341

---

1. Rosa counted mittens, skis, and jackets.

___ ___ ___

___ ___ ___

\+ ___ ___ ___
_____

Rosa counted _____ items in all.

___ ___ ___

---

2. Jerry counted boots, pants, and skis.

___ ___ ___

___ ___ ___

\+ ___ ___ ___
_____

Jerry counted _____ items in all.

___ ___ ___

---

3. Susan counted hats, jackets, and skis.

___ ___ ___

___ ___ ___

\+ ___ ___ ___
_____

Susan counted _____ items in all.

___ ___ ___

---

4. Carl counted mittens, pants, and skis.

___ ___ ___

___ ___ ___

\+ ___ ___ ___
_____

Carl counted _____ items in all.

___ ___ ___

---

5. Carolyn counted hats, skis, and boots.

___ ___ ___

___ ___ ___

\+ ___ ___ ___
_____

Carolyn counted _____ items in all.

___ ___ ___

---

6. Dean counted mittens, boots, and pants.

___ ___ ___

___ ___ ___

\+ ___ ___ ___
_____

Dean counted _____ items in all.

___ ___ ___

Practice 24

Write the number that is 3 more.

1.	2.	3.	4.	5.	6.
13, ___	70, ___	18, ___	94, ___	62, ___	47, ___

Write the number that is 3 less.

7.	8.	9.	10.	11.	12.
90, ___	71, ___	14, ___	57, ___	32, ___	89, ___

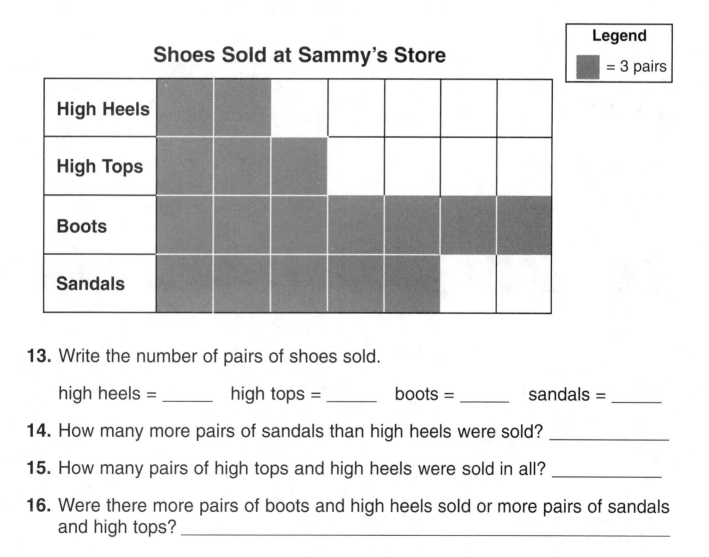

Shoes Sold at Sammy's Store

Legend
= 3 pairs

13. Write the number of pairs of shoes sold.

    high heels = _____   high tops = _____   boots = _____   sandals = _____

14. How many more pairs of sandals than high heels were sold? _____

15. How many pairs of high tops and high heels were sold in all? _____

16. Were there more pairs of boots and high heels sold or more pairs of sandals and high tops? _____

Practice 25

Rewrite each number in standard form.

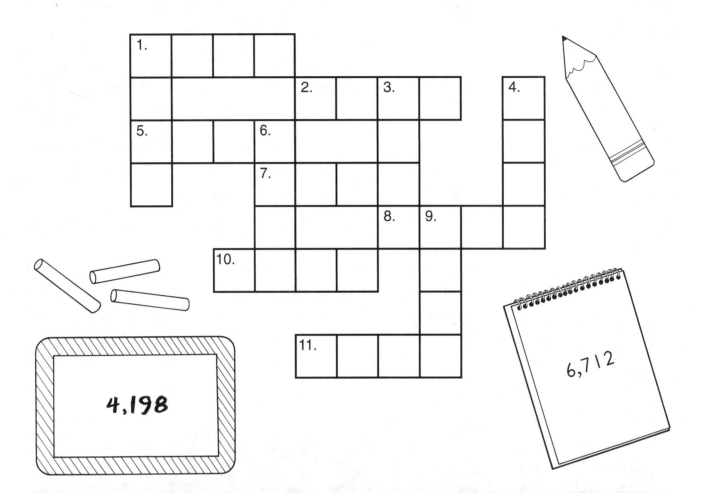

4,198

6,712

	Across		Down
1.	4 thousands + 9 hundreds + 5 tens + 8 ones	**1.**	4 thousands + 1 hundred + 3 tens + 4 ones
2.	7 thousands + 3 hundreds + 5 tens + 2 ones	**3.**	5 thousands + 8 hundreds + 6 tens + 5 ones
5.	3 thousands + 8 hundreds + 3 tens + 9 ones	**4.**	7 thousands + 6 hundreds + 4 tens + 7 ones
7.	8 thousands + 7 hundreds + 8 tens + 6 ones	**6.**	9 thousands + 8 hundreds + 1 ten + 0 ones
8.	5 thousands + 1 hundreds + 8 tens + 7 ones	**9.**	1 thousand + 2 hundreds + 5 tens + 1 one
10.	4 thousands + 0 hundreds + 1 ten + 8 ones		
11.	7 thousands + 9 hundreds + 5 tens + 1 one		

Practice 26

Number of Desserts Sold				
ice cream	cupcakes	cake slices	pie slices	cookies
4,643	5,049	1,119	1,228	2,843

Write the math problem. Write the correct sign (+ or −) in the box.

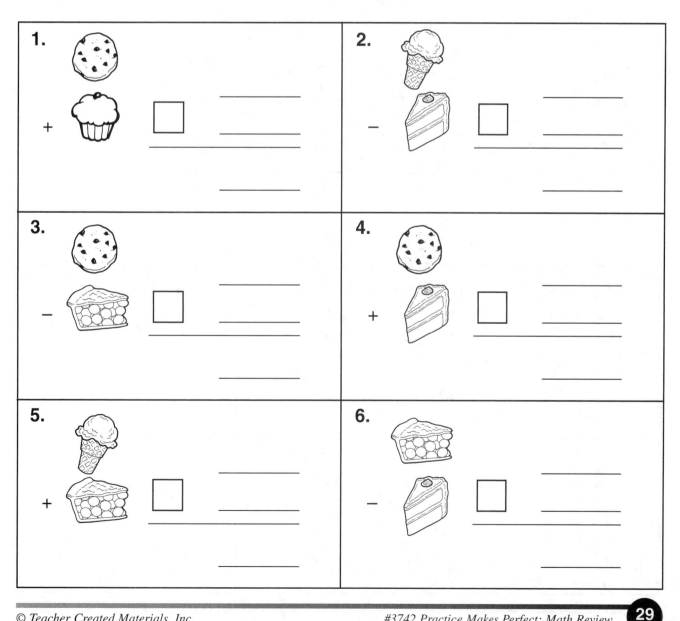

Practice 27

Circle the items to show the fraction. Complete the sentence.

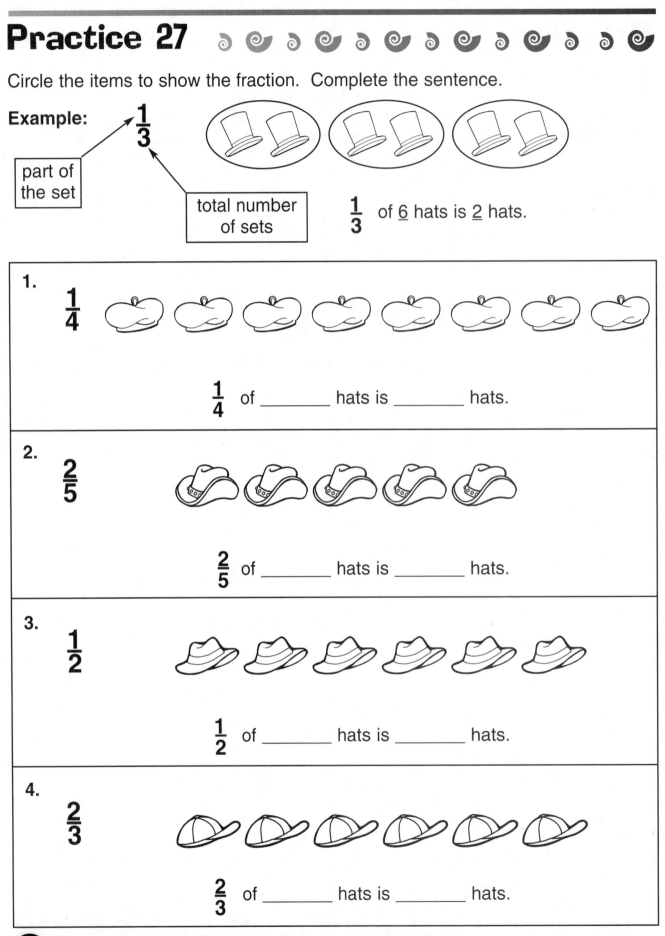

Example: $\frac{1}{3}$

part of the set

total number of sets

$\frac{1}{3}$ of <u>6</u> hats is <u>2</u> hats.

1. $\frac{1}{4}$

$\frac{1}{4}$ of _____ hats is _____ hats.

2. $\frac{2}{5}$

$\frac{2}{5}$ of _____ hats is _____ hats.

3. $\frac{1}{2}$

$\frac{1}{2}$ of _____ hats is _____ hats.

4. $\frac{2}{3}$

$\frac{2}{3}$ of _____ hats is _____ hats.

Practice 28

Solve each problem.

1. Circle 4 equal sets.

There are _____ in each set.

Each set is _____ of the whole.

2. Circle 3 equal sets.

There are _____ in each set.

Each set is _____ of the whole.

3. Circle 6 equal sets.

There are _____ in each set.

Each set is _____ of the whole.

4. Circle 2 equal sets.

There are _____ in each set.

Each set is _____ of the whole.

5. Circle 1 set.

There are _____ in each set.

Each set is _____ of the whole.

6. Shade each circle to show the fraction.

$\frac{1}{8}$ $\frac{1}{5}$ $\frac{1}{4}$ $\frac{1}{3}$

Draw a line under the smallest fraction. Make an X on the largest fraction.

Practice 29

1. Divide the cookies into 2 sets.

Each set is 1/2 of the whole.
1/2 of 12 = _____ cookies

2. Divide the cookies into 3 sets.

Each set is 1/3 of the whole.
1/3 of 12 = _____ cookies

3. Divide the cookies into 4 sets.

Each set is 1/4 of the whole.
1/4 of 12 = _____ cookies

4. Divide the cookies into 6 sets.

Each set is 1/6 of the whole.
1/6 of 12 = _____ cookies

5. Phil ate 3 cookies. Pam ate 1/2 of a dozen cookies. Who ate more cookies?

Phil   Pam

6. Wilma ate 2 cookies. Fred ate 1/4 of a dozen cookies. Who ate more cookies?

Wilma   Fred

7. Grace ate 4 cookies. Tony ate 1/6 of a dozen cookies. Who ate more cookies?

Grace   Tony

8. Raul ate 6 cookies. Amelia ate 1/3 of a dozen cookies. Who ate more cookies?

Raul   Amelia

Practice 30

Draw the picture. Write the addition problems and the multiplication problem for each set of items.

Example: Draw 2 circles. Draw 3 stars in each circle.

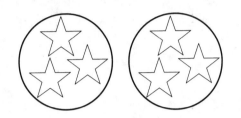

Add:

3 (stars) + 3 (stars) = 6 stars

Multiply:

2 (circles) x 3 (stars in each circle) = 6 stars

---

1. Draw 3 circles. Draw 3 hearts in each circle.

Add: _____ (hearts) + _____ (hearts) + _____ (hearts) = _____ hearts
Multiply: _____ (circles) x _____ (hearts in each circle) = _____ hearts

---

2. Draw 4 circles. Draw 2 fish in each circle.

Add: _____ (fish) + _____ (fish) + _____ (fish) + _____ (fish) = _____ fish
Multiply: _____ (circles) x _____ (fish in each circle) = _____ fish

---

3. Draw a circle. Draw 5 triangles in the circle.

Add: _____ (triangles) + __0__ = _____ triangles
Multiply: _____ (circle) x _____ (triangles in each circle) = _____ triangles

---

Practice 31 ꝺ ꙮ ꝺ ꙮ ꝺ ꙮ ꝺ ꙮ ꝺ ꙮ ꝺ ꙮ ꝺ ꙮ

When multiplying a number by 0, the answer is always 0. Solve the problems.

1.	2.	3.	4.	5.
0 x 2 = ____	8 x 0 = ____	0 x 7 = ____	10 x 0 = ____	9 x 0 = ____
2 x 0 = ____	0 x 8 = ____	7 x 0 = ____	0 x 10 = ____	0 x 9 = ____
6.	7.	8.	9.	10.
6 x 0 = ____	1 x 0 = ____	5 x 0 = ____	12 x 0 = ____	3 x 0 = ____
0 x 6 = ____	0 x 1 = ____	0 x 5 = ____	0 x 12 = ____	3 x 0 = ____

Draw a picture. Write the multiplication problem.

11. Brynn had 3 cookie sheets with 0 cookies on each sheet.	**12.** Flynn had 0 hula hoops in 5 groups.
13. Richie had 0 tokens in 8 groups.	**14.** Ivy had 4 pots with 0 plants in each pot.
15. Carrie had 0 puppies in 10 groups.	**16.** Mason had 8 cars with 0 tires on each car.

Practice 32 ୭ ☙ ୭ ☙ ☙ ୭ ☙ ☙ ☙ ☙ ☙ ☙ ୭ ☙ ☙ ୭ ☙

Write the multiplication problem and then solve each word problem.

1. Carrie brought 4 packs of gum with 2 pieces of gum in each pack. How many pieces of gum are there in all? There are _____ pieces of gum in all.	**2.** Nick bought 5 bags of apples with 2 apples in each bag. How many apples are there in all? There are _____ apples in all.
3. Tim bought 6 ice cream cones with 2 scoops of ice cream on each cone. How many scoops of ice cream are there in all? There are _____ scoops of ice cream in all.	**4.** Duaa bought 10 pairs of shoes with 2 shoes in each pair. How many shoes did Duaa buy in all? Duaa bought _____ shoes in all.
5. Sean picked 3 clovers with 4 leaves on each clover. How many leaves are there in all? There are _____ leaves in all.	**6.** Isabelle bought 1 pizza and cut it into 8 slices. How many slices of pizza are in all? There are _____ slices of pizza in all.

Multiply to solve each problem.

7. $1 \times 9 =$ _____   **8.** $3 \times 3 =$ _____   **9.** $5 \times 3 =$ _____

10. $6 \times 1 =$ _____   **11.** $4 \times 4 =$ _____   **12.** $8 \times 2 =$ _____

Practice 33

Draw the picture. Write the word problem and division problem for each set of items.

Example: 9 stars in 3 groups

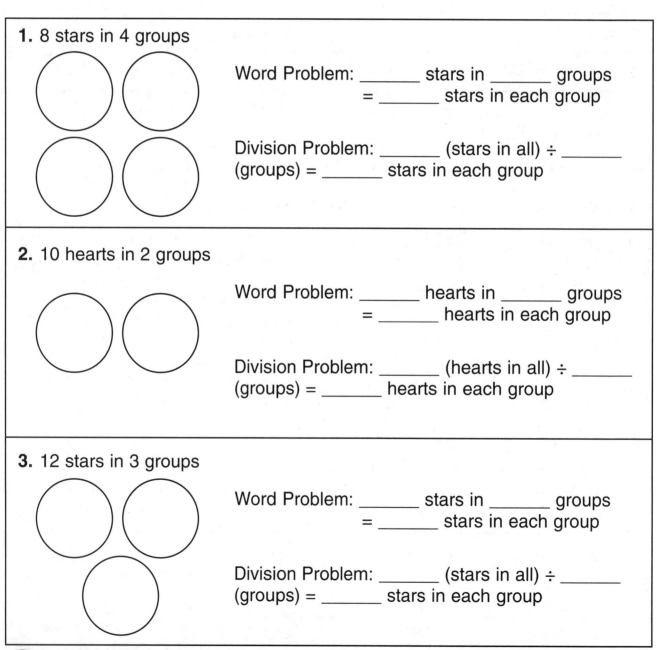

Word Problem

9 (stars in all) in 3 groups = 3 stars in each group

Division Problem

9 (stars in all) ÷ 3 (groups) = 3 stars in each group

1. 8 stars in 4 groups

Word Problem: _____ stars in _____ groups
= _____ stars in each group

Division Problem: _____ (stars in all) ÷ _____ (groups) = _____ stars in each group

2. 10 hearts in 2 groups

Word Problem: _____ hearts in _____ groups
= _____ hearts in each group

Division Problem: _____ (hearts in all) ÷ _____ (groups) = _____ hearts in each group

3. 12 stars in 3 groups

Word Problem: _____ stars in _____ groups
= _____ stars in each group

Division Problem: _____ (stars in all) ÷ _____ (groups) = _____ stars in each group

Practice 34 ৯ ৫ ৯ ৫ ৫ ৯ ৫ ৯ ৫ ৯ ৫ ৯ ৫

When multiplying or dividing a number by 1, the answer is always the same as the other number. Solve each problem.

1.	2.	3.	4.	5.
1 x 2 = ____	3 x 1 = ____	1 x 4 = ____	5 x 1 = ____	6 x 1 = ____
2 x 1 = ____	1 x 3 = ____	4 x 1 = ____	1 x 5 = ____	1 x 6 = ____
6.	7.	8.	9.	10.
2 ÷ 1 = ____	3 ÷ 1 = ____	5 ÷ 1 = ____	7 ÷ 1 = ____	9 ÷ 1 = ____
1 ÷ 1 = ____	4 ÷ 1 = ____	6 ÷ 1 = ____	8 ÷ 1 = ____	10 ÷ 1 = ____

Circle whether you would multiply or divide to solve the problem.

11. Sebastian has 10 hats in 2 groups. How many hats are in each group? multiply divide	12. Eartha has 4 coin purses. Eartha has 3 coins in each purse. How many coins are there in all? multiply divide
13. Santos had 5 flowers. Each flower has 2 buds. How many buds are there in all? multiply divide	14. Mariska put 9 books in 3 stacks. How many books are in each stack? multiply divide
15. Tucker has 4 wheels on 2 bikes. How many wheels are on each bike? multiply divide	16. Gertie has 2 boxes. In each box there are 6 necklaces. How many necklaces are there in all? multiply divide

Practice 35

1. Circle the numbers used when counting by 3.

0   1   2   3   4   5   6   7   8   9   10   11   12   13

14   15   16   17   18   19   20   21   22   23   24   25   26

2. Complete the table.

x 3	0	1	2	3	4	5	6	7	8	9	10
	0			9				21			30

3. Complete the table.

÷ 3	30	27	24	21	18	15	12	9	6	3	0
	10			7			4			1	

Write the equation and then solve each problem. Use the charts to help you.

4. Crystal has 3 vases with 4 flowers in each vase. How many flowers are there in all? There are _____ flowers in all.	**5.** Lionel has 12 pencils in 4 pencil boxes. How many pencils are in each box? There are _____ pencils in each box.
6. Chloris has 18 seashells in 6 boxes. How many seashells are in each box? There are _____ seashells in each box.	**7.** Wesley has 7 baskets with 3 berries in each basket. How many berries are there in all? There are _____ berries in all.

Practice 36 ɔ ❧ ❧ ɔ ❧ ɔ ❧ ❧ ɔ ❧ ❧ ɔ ❧ ❧ ɔ ɔ ❧

Cross out the information that is not needed to solve the problem. Circle the math operation (+, −, x, or ÷) needed to solve the problem.

1. Jade has 6 friends. Each friend has 2 cookies. The friends like the cookies. How many cookies are there in all?
 	
  +   −   x   ÷
	2. Gus has 5 dogs and 4 cats. The animals make a lot of noise. Gus has 3 birds. How many animals are there in all?
	
	  + − x ÷

3. Katherine has 5 trees. She picked 9 apples. She made 3 pies with the apples. How many apples are in each pie?
 	
  +   −   x   ÷
	4. Jacob had 20 masks. He sold 12 of them at the fair. His favorite mask is green. How many masks are left?
	
	  + − x ÷

5. Sada picked up 9 snails and 14 slugs. She found them outside. How many bugs are there in all?
 	
  +   −   x   ÷
	6. Leland counted 8 people wearing 2 gloves each. Leland hates to wear gloves. How many gloves are there in all?
	
	  + − x ÷

7. Hayley made 18 brownies. She sold 13 of them. The brownies had nuts in them. How many brownies are left?
 	
  +   −   x   ÷
	8. Damon made 15 pancakes. He put the pancakes on 5 plates. Damon's mom doesn't like pancakes. How many pancakes are on each plate?
	
	  + − x ÷

Test Practice 1

Fill in the correct answer bubble.

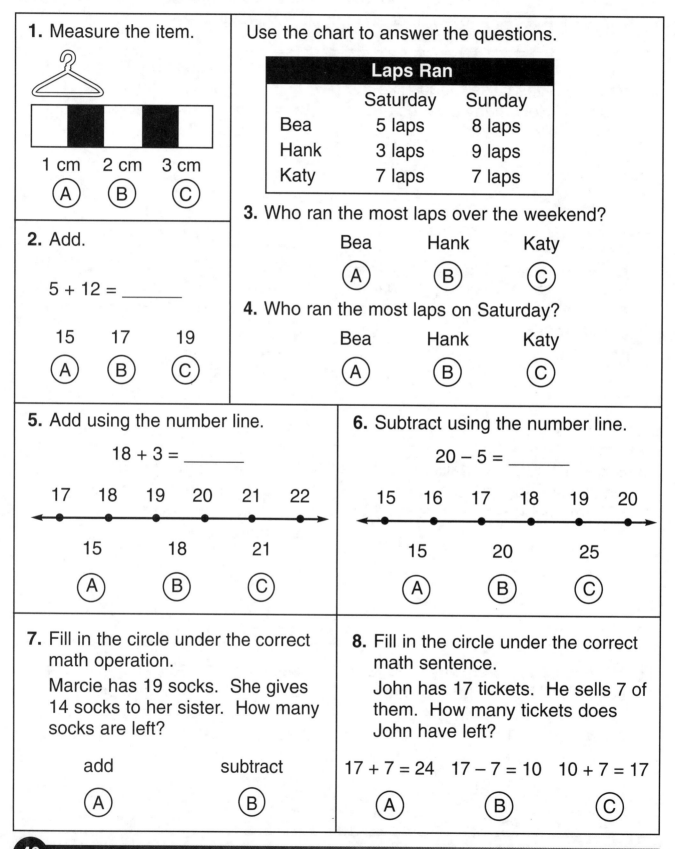

1. Measure the item.

1 cm 2 cm 3 cm
(A)   (B)   (C)

2. Add.

5 + 12 = _____

15   17   19
(A)   (B)   (C)

Use the chart to answer the questions.

Laps Ran		
	Saturday	Sunday
Bea	5 laps	8 laps
Hank	3 laps	9 laps
Katy	7 laps	7 laps

3. Who ran the most laps over the weekend?

Bea   Hank   Katy
(A)   (B)   (C)

4. Who ran the most laps on Saturday?

Bea   Hank   Katy
(A)   (B)   (C)

5. Add using the number line.

18 + 3 = _____

17   18   19   20   21   22

15   18   21
(A)   (B)   (C)

6. Subtract using the number line.

20 – 5 = _____

15   16   17   18   19   20

15   20   25
(A)   (B)   (C)

7. Fill in the circle under the correct math operation.

Marcie has 19 socks. She gives 14 socks to her sister. How many socks are left?

add   subtract
(A)   (B)

8. Fill in the circle under the correct math sentence.

John has 17 tickets. He sells 7 of them. How many tickets does John have left?

17 + 7 = 24   17 – 7 = 10   10 + 7 = 17
(A)   (B)   (C)

#3742 Practice Makes Perfect: Math Review   © *Teacher Created Materials, Inc.*

Test Practice 2 ✺ ✺ ✺ ✺ ✺ ✺ ✺ ✺ ✺ ✺ ✺ ✺

Fill in the correct answer bubble.

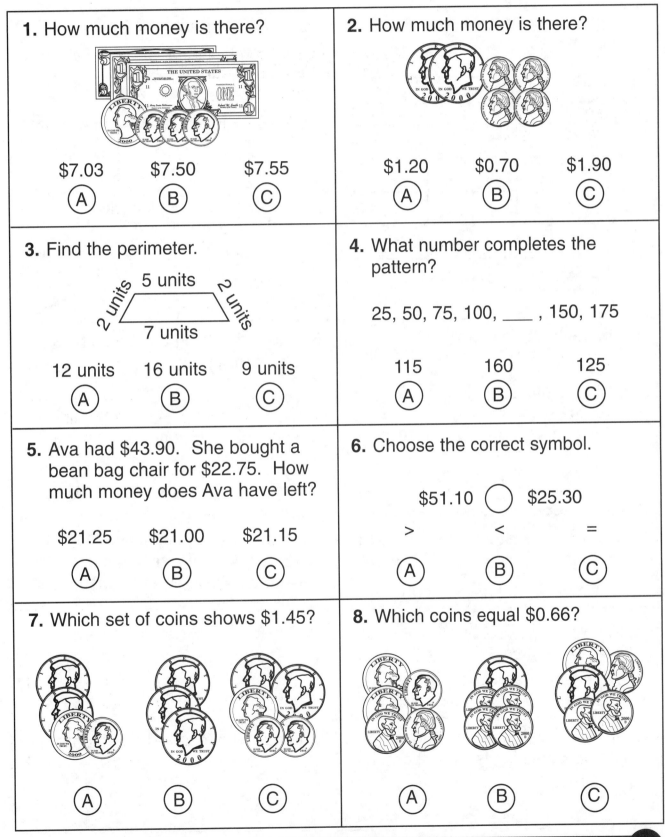

1. How much money is there?

$7.03 $7.50 $7.55
(A)   (B)   (C)

2. How much money is there?

$1.20 $0.70 $1.90
(A)   (B)   (C)

3. Find the perimeter.

5 units
2 units   2 units
7 units

12 units   16 units   9 units
(A)   (B)   (C)

4. What number completes the pattern?

25, 50, 75, 100, ___ , 150, 175

115   160   125
(A)   (B)   (C)

5. Ava had $43.90. She bought a bean bag chair for $22.75. How much money does Ava have left?

$21.25 $21.00 $21.15
(A)   (B)   (C)

6. Choose the correct symbol.

$51.10 ◯ $25.30

>   <   =
(A)   (B)   (C)

7. Which set of coins shows $1.45?

(A)   (B)   (C)

8. Which coins equal $0.66?

(A)   (B)   (C)

#3742 Practice Makes Perfect: Math Review

Test Practice 3

Fill in the correct answer bubble.

Use the calendar to answer the questions 1–3.

JANUARY						
Sunday	Monday	Tuesday	Wednesday	Thursday	Friday	Saturday
		1	2	3	4	5
6	7	8	9	10	11	12
13	14	15	16	17	18	19
20	21	22	23	24	25	26
27	28	29	30	31		

1. What is the name of the month?

December (A)   January (B)   February (C)

2. How many days are in this month?

30 (A)   31 (B)   32 (C)

3. What day of the week is the 22nd?

Tuesday (A)   Thursday (B)   Saturday (C)

4. Complete the pattern.

15, 30, _____ , 60, 75

45 (A)   35 (B)   90 (C)

5. About how long could you hold your breath?

10 seconds (A)   10 minutes (B)   10 hours (C)

6. It is 3:45. What time will it be in 15 minutes?

(A)   (B)   (C)

7. A new bus leaves every 15 minutes. If a bus left at 7:15, what time will the next bus leave?

7:00 (A)   7:45 (B)   7:30 (C)

8. At the circus there were 30 big cats, 40 bears, 10 birds, and 10 elephants. How many animals were there in all?

90 (A)   100 (B)   80 (C)

Test Practice 4 ͻ ☙ ͻ ☙ ͻ ☙ ͻ ☙ ͻ ☙ ͻ ͻ ☙

Fill in the correct answer bubble.

1. Write the number in standard form. 5 tens 8 ones 85 13 58 Ⓐ Ⓑ Ⓒ	**2.** Write the number in standard form. 3,000 + 100 + 70 + 6 3,761 3,167 3,176 Ⓐ Ⓑ Ⓒ	**3.** Write the number in standard form. two hundred fifteen 20,015 215 250 Ⓐ Ⓑ Ⓒ

4. Solve the riddle.

I have a 3 in the hundreds place and a 6 in the ones place. Which number am I?

631   316   136
Ⓐ   Ⓑ   Ⓒ

5. Solve the problem.

Ben had 22 hamsters. He bought 35 more. How many hamsters does Ben now have?

20 + 35   22 + 35   57 + 22
Ⓐ   Ⓑ   Ⓒ

6. Solve the problem.

Dusty had 483 napkins. She folded 161 of the napkins to look like swans. How many napkins does Dusty have left to fold?

320   322   344
Ⓐ   Ⓑ   Ⓒ

7. Solve the problem.

Christopher counted 1,349 toothpicks and 4,215 beans. How many items did Christopher count?

5,564   5,654   6,564
Ⓐ   Ⓑ   Ⓒ

8. Andrea collected 127 bottles and 306 cans. How many items did Andrea collect? 344 433 343 Ⓐ Ⓑ Ⓒ	**9.** What number is 20 more than 625? 605 650 645 Ⓐ Ⓑ Ⓒ	**10.** Choose the correct symbol. 2,490 ◯ 2,940 > < = Ⓐ Ⓑ Ⓒ

#3742 Practice Makes Perfect: Math Review

Test Practice 5

Fill in the correct answer bubble.

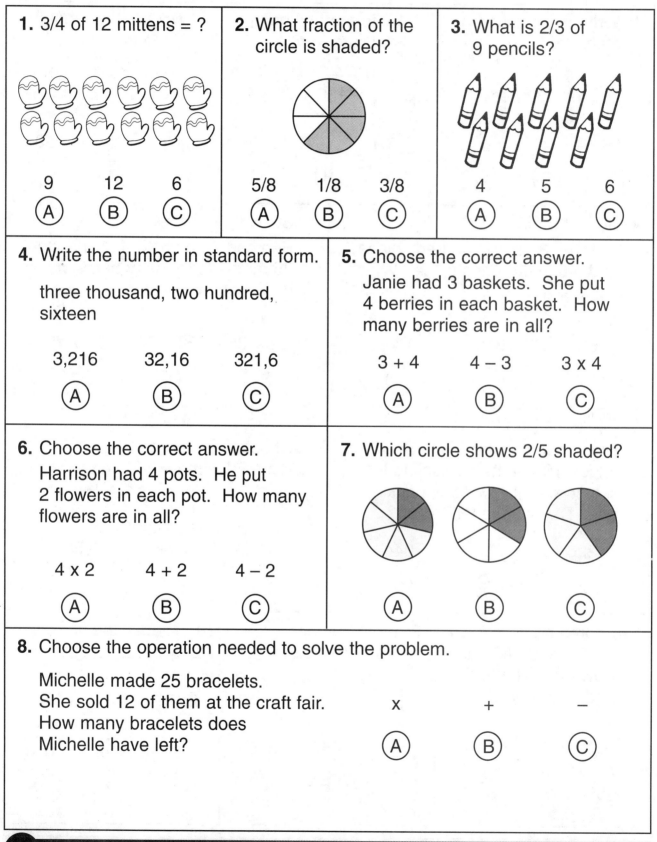

1. 3/4 of 12 mittens = ?

9	12	6
Ⓐ	Ⓑ	Ⓒ

2. What fraction of the circle is shaded?

5/8	1/8	3/8
Ⓐ	Ⓑ	Ⓒ

3. What is 2/3 of 9 pencils?

4	5	6
Ⓐ	Ⓑ	Ⓒ

4. Write the number in standard form.

three thousand, two hundred, sixteen

3,216	32,16	321,6
Ⓐ	Ⓑ	Ⓒ

5. Choose the correct answer.

Janie had 3 baskets. She put 4 berries in each basket. How many berries are in all?

3 + 4	4 – 3	3 x 4
Ⓐ	Ⓑ	Ⓒ

6. Choose the correct answer.

Harrison had 4 pots. He put 2 flowers in each pot. How many flowers are in all?

4 x 2	4 + 2	4 – 2
Ⓐ	Ⓑ	Ⓒ

7. Which circle shows 2/5 shaded?

Ⓐ	Ⓑ	Ⓒ

8. Choose the operation needed to solve the problem.

Michelle made 25 bracelets. She sold 12 of them at the craft fair. How many bracelets does Michelle have left?

x	+	–
Ⓐ	Ⓑ	Ⓒ

Test Practice 6

Fill in the correct answer bubble.

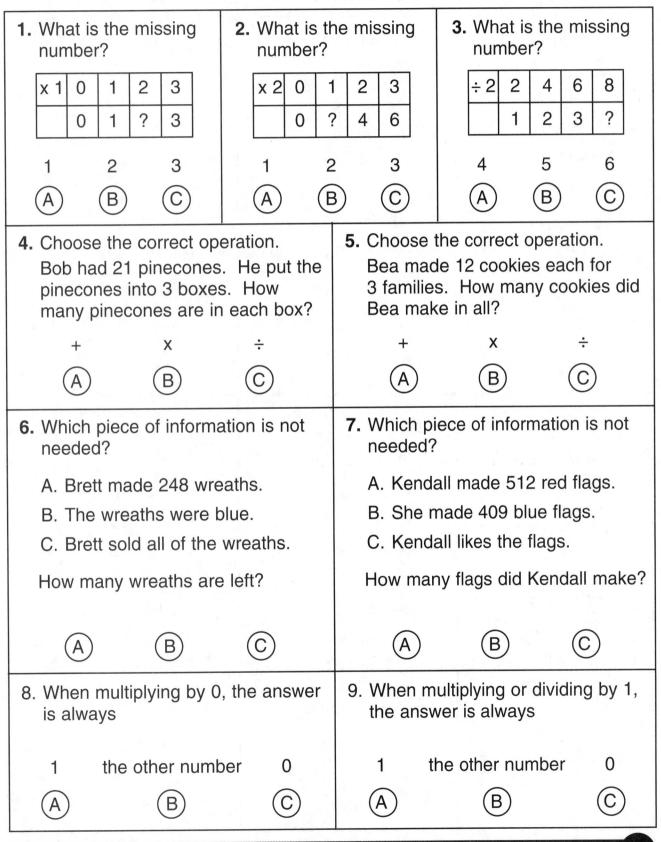

1. What is the missing number?

x 1	0	1	2	3
	0	1	?	3

   1   2   3
  Ⓐ   Ⓑ   Ⓒ

2. What is the missing number?

x 2	0	1	2	3
	0	?	4	6

   1   2   3
  Ⓐ   Ⓑ   Ⓒ

3. What is the missing number?

÷ 2	2	4	6	8
	1	2	3	?

   4   5   6
  Ⓐ   Ⓑ   Ⓒ

4. Choose the correct operation.

Bob had 21 pinecones. He put the pinecones into 3 boxes. How many pinecones are in each box?

   +   x   ÷
  Ⓐ   Ⓑ   Ⓒ

5. Choose the correct operation.

Bea made 12 cookies each for 3 families. How many cookies did Bea make in all?

   +   x   ÷
  Ⓐ   Ⓑ   Ⓒ

6. Which piece of information is not needed?

A. Brett made 248 wreaths.

B. The wreaths were blue.

C. Brett sold all of the wreaths.

How many wreaths are left?

  Ⓐ   Ⓑ   Ⓒ

7. Which piece of information is not needed?

A. Kendall made 512 red flags.

B. She made 409 blue flags.

C. Kendall likes the flags.

How many flags did Kendall make?

  Ⓐ   Ⓑ   Ⓒ

8. When multiplying by 0, the answer is always

   1   the other number   0
  Ⓐ   Ⓑ   Ⓒ

9. When multiplying or dividing by 1, the answer is always

   1   the other number   0
  Ⓐ   Ⓑ   Ⓒ

Answer Sheet

Test Practice 1

1. (A) (B) (C)
2. (A) (B) (C)
3. (A) (B) (C)
4. (A) (B) (C)
5. (A) (B) (C)
6. (A) (B) (C)
7. (A) (B)
8. (A) (B) (C)

Test Practice 2

1. (A) (B) (C)
2. (A) (B) (C)
3. (A) (B) (C)
4. (A) (B) (C)
5. (A) (B) (C)
6. (A) (B) (C)
7. (A) (B) (C)
8. (A) (B) (C)

Test Practice 3

1. (A) (B) (C)
2. (A) (B) (C)
3. (A) (B) (C)
4. (A) (B) (C)
5. (A) (B) (C)
6. (A) (B) (C)
7. (A) (B) (C)
8. (A) (B) (C)

Test Practice 4

1. (A) (B) (C)
2. (A) (B) (C)
3. (A) (B) (C)
4. (A) (B) (C)
5. (A) (B) (C)
6. (A) (B) (C)
7. (A) (B) (C)
8. (A) (B) (C)
9. (A) (B) (C)
10. (A) (B) (C)

Test Practice 5

1. (A) (B) (C)
2. (A) (B) (C)
3. (A) (B) (C)
4. (A) (B) (C)
5. (A) (B) (C)
6. (A) (B) (C)
7. (A) (B) (C)
8. (A) (B) (C)

Test Practice 6

1. (A) (B) (C)
2. (A) (B) (C)
3. (A) (B) (C)
4. (A) (B) (C)
5. (A) (B) (C)
6. (A) (B) (C)
7. (A) (B) (C)
8. (A) (B) (C)
9. (A) (B) (C)

Answer Key

Page 4
1. 9
2. 14
3. 9
4. 10
5. 18
6. 16
7. +
8. +
9. −
10. −
11. +
12. −

Page 5
1. 18
2. 11
3. 13
4. 10
5. 9
6. 14
7. 17
8. 12
9. 8
Message: Stars and Stripes

Page 6
1. +
2. −
3. +
4. −
5. −
6. +
7. in all, 5 + 5 = 10
8. have left, 9 − 6 = 3
9. difference, 7 − 2 = 5
10. all together, 1 + 8 = 9
11. fewer, 7 − 5 = 2

Page 7
1. 96
2. 50
3. 90
4. 62
5. 50
6. 94
7. Ken
8. Forest
9. 80
10. 84

Page 8
1. 5
2. 3
3. 6
4. 4
5. 8
6. 2
7. 4
8. 6

Page 9
1. 4, 4
2. 8, 8
3. 5, 5
4. 3, 3
5. B
6. C
7. C
8. B

Page 10
1. 2 + 4 + 4 = 10 units
2. 2 + 2 + 4 + 4 = 12 units
3. 3 + 3 + 3 + 3 = 12 units
4. 2 + 2 + 2 + 2 + 2 = 10 units
5. 1 + 1 + 1 + 1 + 1 + 1 = 6 units
6. 2 + 2 + 2 + 2 = 8 units
7. hexagon
8. rhombus

Page 11
1. 1 dime, 2 nickels, 2 pennies
2. 1 quarter, 2 dimes, 1 nickel, 1 penny
3. 1 nickel, 4 pennies
4. 2 quarters, 3 dimes
5. 2 quarters, 2 dimes, 1 penny

Page 12
1. 75, 125, 150, 175, 225
2. $1.25
3. $0.75
4. $0.25
5. $1.50
6. $1.00
7. $1.75

Page 13
1. 75¢
2. 17¢
3. 60¢
4. 30¢
5. yes
6. no
7. yes
8. no
9. yes
10. no

Page 14
1. $5.75
2. $3.37
3. $11.70
4. $4.50
5. $25.03
6. $6.29
7. $3.37, $4.50, $5.75, $6.29, $11.70, $25.03

Page 15
1.

$79.50	$28.13
− $51.37	+ $51.37
$28.13	$79.50

Theo has $28.13 left.

2.

$13.10	$ 2.04
− $11.06	+ $11.06
$ 2.04	$13.10

Queena has $2.04 left.

3.

$11.75	$ 1.27
− $10.48	+ $10.48
$ 1.27	$11.75

Thomas has $1.27 left.

Page 16
1.

$38.95	$ 8.29
− $30.66	+ $30.66
$ 8.29	$38.95

Maurine has $8.29 left.

2.

$86.93	$62.14
− $24.79	+ $24.79
$62.14	$86.93

Monty has $62.14 left.

3. Answers will vary.
4. >
5. <
6. <

Page 17

Sunday	Monday	Tuesday	Wednesday	Thursday	Friday	Saturday
						1
2	3	4	5	6	7	8
9	10	11	12	13	14	15
16	17	18	19	20	21	22
23	24	25	26	27	28	29
30	31					

1. December
2. 31
3. Monday
4. Saturday
5. Wednesday
6. Tuesday
7. Monday
8. Sunday

Page 18
1. 8:30, 8:45, 9:15, 9:30, 9:45
2. 2:00, 2:15, 2:45, 3:00, 3:15
3. 3:30, 3:15, 3:00, 2:45, 2:30
4. 11:15, 11:00, 10:45, 10:30, 10:15
5. 9:00
6. 1:30
7. 4:45
8. 6:45
9. 3:30
10. 5:00

Page 19
1. a quarter past 10
2. half past 8
3. a quarter till 4
4. half past 7
5. half past 8
6. a quarter til 6
7. a quarter til 7
8. half past 1

Page 20
1. sleeping bag 1, car 4, bus 70
2. large tub 30, milk jug 1, sink 5
3. skis 5, bat 3, hanger 1
4. bucket 1,000, bag 100, pocket 10
5. man 150, baby 10, boy 50
6. cupcake 20, turkey 200, egg 2
7. flower 1, rake 5, plant 50
8. tree 25, grapes 10, apple 1
9. city 12, igloo 1, house 3

Page 21
1. 25 + 40 = 65, Reba counted 65 nuts in all.
2. 50 + 31 = 81, Jordan counted 81 nuts in all.
3. 86 − 33 = 53, There are 53 walnuts left.
4. 60 + 14 = 74, Patsy counted 74 nuts in all.
5. 95 − 51 = 44, There are 44 peanuts left.
6. 73 − 22 = 51, There are 51 pistachios left.

Page 22
1. 334 Orange Street
2. 788 Pecan Avenue
3. 443 Walnut Lane
4. 911 Mandarin Lane
5. 619 Lemon Street
6. 682 Coconut Boulevard
7. Answers will vary.

Page 23
Lisa's Pet: Cross off clue #3, Ralph

Corey's Pet: Cross off clue #1, Buster

Ted's Pet: Cross off clue #2, Dudley

Anna's Pet: Cross off clue #3, Oscar

Answer Key

Page 24
1. 102
2. 999
3. 753
4. 674
5. 365
6. 482

Page 25
1. 104 < 675, Graciela
2. 176 < 425, Nancy
3. 584 > 324, Brittany
4. 518 > 219, Kelly
5. 613 < 841, Omari
6. 297 < 669, Ken
7. 234 < 589, Joe
8. 597 > 210, Gwen

Page 26
1. 403 + 230 + 154 = 787, Rosa counted 787 items in all.
2. 125 + 341 + 230 = 696, Jerry counted 696 items in all.
3. 512 + 154 + 230 = 896, Susan counted 896 items in all.
4. 403 + 341 + 230 = 974, Carl counted 974 items in all.
5. 512 + 230 + 125 = 867, Carolyn counted 867 items in all.
6. 403 + 125 + 341 = 869, Dean counted 869 items in all.

Page 27
1. 16
2. 73
3. 21
4. 97
5. 65
6. 50
7. 87
8. 68´
9. 11
10. 54
11. 29
12. 86
13. 6, 9, 21, 15
14. 9
15. 15
16. boots and high heels

Page 28
Across
1. 4,958
2. 7,352
5. 3,839
7. 8,786
8. 5,187
10. 4,018
11. 7,951

Down
1. 4,134
3. 5,865
4. 7,647
6. 9,810
9. 1,251

Page 29
1. 2,843 + 5,049 = 7,892
2. 4,643 − 1,119 = 3,524
3. 2,843 − 1,228 = 1,615
4. 2,843 + 1,119 = 3,962
5. 4,643 + 1,228 = 5,871
6. 1,228 − 1,119 = 109

Page 30
1. 1/4 of 8 hats is 2 hats.
2. 2/5 of 5 hats is 2 hats.
3. 1/2 of 6 hats is 3 hats.
4. 2/3 of 6 hats is 4 hats.

Page 31
1. 3, 1/4
2. 4, 1/3
3. 2, 1/6
4. 6, 1/2
5. 12, 1
6. Check to see that the correct amounts are shaded in each circle.
smallest = 1/8,
largest = 1/3

Page 32
1. 6
2. 4
3. 3
4. 2
5. Pam
6. Fred
7. Grace
8. Raul

Page 33
1. 3 + 3 + 3 = 9, 3 x 3 = 9
2. 2 + 2 + 2 + 2 = 8, 4 x 2 = 8
3. 5 + 0 = 5, 1 x 5 = 5

Page 34
1–10. All answers are 0.
11. 3 x 0 = 0
12. 0 x 5 = 0
13. 0 x 8 = 0
14. 4 x 0 = 0
15. 10 x 0 = 0
16. 8 x 0 = 0

Page 35
1. 4 x 2 = 8
2. 5 x 2 = 10
3. 6 x 2 = 12
4. 10 x 2 = 20
5. 3 x 4 = 12
6. 1 x 8 = 8
7. 9
8. 9
9. 15
10. 6
11. 16
12. 16

Page 36
1. Word Problem: 8 stars in 4 groups = 2 stars in each group. Division Problem: 8 ÷ 4 = 2
2. Word Problem: 10 hearts in 2 groups = 5 hearts in each group. Division Problem: 10 ÷ 2 = 5
3. Word Problem: 12 stars in 3 groups = 4 stars in each group. Division Problem: 12 ÷ 3 = 4

Page 37
1. 2, 2
2. 3, 3
3. 4, 4
4. 5, 5
5. 6, 6
6. 2, 1
7. 3, 4
8. 5, 6

9. 7, 8
10. 9, 10
11. divide
12. multiply
13. multiply
14. divide
15. divide
16. multiply

Page 38
1. 0, 3, 6, 9, 12, 15, 18, 21, 24
2. 3, 6, 12, 15, 18, 24, 27
3. 9, 8, 6, 5, 3, 2, 0
4. 3 x 4 = 12
5. 12 ÷ 4 = 3
6. 18 ÷ 6 = 3
7. 7 x 3 = 21

Page 39
1. Cross out: The friends like the cookies, x or +
2. Cross out: The animals make a lot of noise, +
3. Cross out: Katherine has 5 trees, ÷
4. Cross out: His favorite mask is green, −
5. Cross out: She found them outside, +
6. Cross out: Leland hates to wear gloves, x or +
7. Cross out: The brownies had nuts in them, −
8. Cross out: Damon's mom doesn't like pancakes, ÷

Page 40
1. B
2. B
3. C
4. C
5. C
6. A
7. B
8. B

Page 41
1. C
2. A
3. B

4. C
5. C
6. A
7. C
8. A

Page 42
1. B
2. B
3. A
4. A
5. A
6. C
7. C
8. A

Page 43
1. C
2. C
3. B
4. B
5. B
6. B
7. A
8. B
9. C
10. B

Page 44
1. A
2. A
3. C
4. A
5. C
6. A
7. C
8. C

Page 45
1. B
2. B
3. A
4. C
5. B
6. B
7. C
8. C
9. B